Kishodai Tokyo Kanku

Organization of the Meteorological System in Japan

Kishodai Tokyo Kanku

Organization of the Meteorological System in Japan

ISBN/EAN: 9783337034795

Printed in Europe, USA, Canada, Australia, Japan

Cover: Foto ©berggeist007 / pixelio.de

More available books at **www.hansebooks.com**

ORGANIZATION

OF THE

METEOROLOGICAL SYSTEM

IN

JAPAN.

CENTRAL METEOROLOGICAL OBSERVATORY
OF JAPAN.
TOKIO.
1893.

PREFACE

This volume has been prepared for the World's Columbian Exposition at Chicago with the purpose of showing our Meteorological System to the visitors from various foreign countries on this unique occasion.

It must be distinctly understood that the volume chiefly refers to the present organization of Meteorological System in Japan.

For the sake of convenience this volume has been compiled in strict accordance with the Circular issued by the International Meteorological Committee on December 31st, 1882.

I express here my sincerest thanks to Mr. N. Baba, assistant meteorologist of the Service of Indications, to whom I have entrusted the task of compiling this volume.

K. KOBAYASHI,

Director of Central Meteorological Observatory.

Tokio, January 1893.

CONTENTS.

ORGANIZATION

OF THE

METEOROLOGICAL SYSTEM

IN

JAPAN.

GENERAL DEFINITION OF THE METEOROLOGICAL SYSTEM.

The Regulations relating to the Meteorological System in Japan have been established by an Imperial Ordinance, the rules for enforcing the same being fixed by the Minister for Home Affairs.

We hereby give Our sanction to the Regulations relating to the Meteorological Observatory and Stations and order the same to be promulgated.

(H. I. M.'s Sign Manual)
(Privy Seal)
The 3rd day, the 8th month, the 20th year of *Meiji*, (1887).

Count ITO HIROBUMI
Minister President of State.

Count YAMAGATA ARITOMO
Minister of State for Home Affairs.

IMPERIAL ORDINANCE No. XLI.—REGULATIONS RELATING TO THE
METEOROLOGICAL OBSERVATORY AND STATIONS.

ART. 1—A Central Observatory is to be founded in Tokio and number of Provincial Stations in suitable locations. The positions of the same shall be determined by the Minister for Home Affairs.

ART. 2—Any persons who desire to erect meteorological stations besides those above mentioned, shall obtain sanction from the Minister for Home Affairs.

ART. 3—The Central Observatory shall be directly controlled by the Minister for Home Affairs and Provincial Stations by the Governors of the Prefectures under the supervision of the Minister. The other stations shall be placed under the supervision of the Governors of the Prefectures.

ART. 4—The expenses of Provincial Stations shall be sustained from the revenues of the respective Prefectures concerned.

ART. 5—The Central Observatory and Provincial Stations shall be in communication with one another in the execution of their business.

ART. 6—The rules for enforcing these regulations shall be fixed by the Minister for Home Affairs.

NOTIFICATION No. V OF THE DEPARTMENT FOR HOME AFFAIRS.

The rules for enforcing the Regulations relating to the Meteorological Observatory and Stations are hereby revised as follows:—

The 4th day, the 5th month, the 25th year of *Meiji*, (1892).

Count SOYESIIIMA TANEOMI
Minister of State for Home Affairs.

The Rules for enforcing the Regulations relating to the Meteorological Observatory and Stations.

ART. 1—The Central Observatory shall control the meteorological matters and investigate the meteorology of the Empire and issue weather forecasts and storm warnings for the Empire.

ART. 2—The Provincial Stations shall make meteorological observations of the locality, investigate climate in the administrative districts of the Prefectures concerned and issue the provincial weather forecasts according to the forecasts issued from the Central Meteorological Observatory.

ART. 3—The Provincial Stations are divided to the 1st and the 2nd order.

The 1st order Stations shall be provided with barometer, thermometer, hygrometer, maximum and minimum thermometers, solar radiation thermometer, terrestrial · radiation thermometer, earth thermometer, anemometer, anemoscope, raingauge, evapometer, sunshine-recorder, seismograph and make hourly observations.

If provided with self-recording instruments, the number of observations may be reduced to six times a day.

The 2nd order Stations shall be provided with barometer, thermometer, hygrometer, maximum and minimum thermometer, anemometer, anemoscope, raingauge, seismograph and make observations six times a day.

ART. 4—When a storm warning is received from the Central Observatory or the weather appears threatening, the Provincial Stations shall make extraordinary observations besides those above required.

ART. 5—The Provincial Stations have to forward the following records to the Central Observatory :—

Meteorological telegrams.
Monthly meteorological report.
Yearly meteorological report.
Yearly report of the business transactions.
Five-yearly meteorological report.
Storm record.
Thunderstorm record.
Earthquake record.
Snow record.
Record relating to animals.
Record relating to plants.
Miscellaneous reports.

ART. 6—According to the provisions of Art. 5 of the Regulations the Provincial Stations shall print monthly, yearly, and 5 yearly meteorological reports, which each station may exchange with the other.

ART. 7—Hours and methods of observations, characters of instruments, and forms and periods of records, shall be determined by the Central Observatory with the approval of the Minister for Home Affairs.

ART. 8—The Minister for Home Affairs may order meteorologists of the Central Observatory to inspect the Provincial Stations.

ART. 9—Any person who desires in accordance with provisions of Art. 2 of the Regulations to erect a meteorological station shall apply for the sanction of the Minister for Home Affairs through the Governor of Prefecture concerned, reporting full particulars on the following matters:

1. Position of the establishment and its topography; (a sketch of the position shall be appended).
2. Full descriptions of the instruments to be used.
3. Mode of maintenance.

ART. 10—The expenses of Provincial Stations mentioned in the Art. 4 of the Regulations shall be defrayed in the case of Hokkaido, from the appropriations of that Administration Board, and in the case of Okinawa from the income of that Prefecture.

SUPPLEMENTARY RULES.

ART. 11—The issuance of the provincial weather forecasts, according to the provisions of Art. 2 shall require, for time being, the sanction of the Minister for Home Affairs.

ART. 12—Cities, towns, villages or private individuals desiring to erect new Storm Signals shall obtain the sanctions of the Governors of the Prefectures concerned.

ART. 13—In case a Governor of Prefecture erect a new Storm Signal or give sanction to its erection, the fact shall be reported to the Central Observatory at least 30 days before the signal is put in operation, with its full particulars and the sketch of its position.

ART. 14—In case any changes are made in connection with Storm Signals, the Governor of Prefecture concerned shall report the fact to the Central Observatory.

ART. 15—In case stated in the Art. 13 and 14 the Central Observatory shall notify the fact throughout the Empire.

NOTIFICATION No. IV OF THE DEPARTMENT FOR HOME AFFAIRS.

In accordance with the Imperial Ordinance No. 41, relating to the Meteorological Observatory and Stations, the positions of Provincial Meteorological Stations are determined as follows:

The 11th day, the 10th Month, the 20th Year of *Meiji*, (1887).

Count YAMAGATA ARITOMO
Minister of State for Home Affairs.

Hakodate Sapporo Nemuro Suttsu Erimo Soya	in Hokkaido cho
Kioto	Kioto fu
Ozaka	Ozaka fu
Yokohama	Kanagawa ken
Kobe	Hiogo ken
Nagasaki Itsugahara	Nagasaki ken
Niigata	Niigata ken
Urawa	Saitama ken
Maibashi	Gumba ken
Choshi	Chiba ken
Mito	Ibaraki ken
Utsunomiya	Tochigi ken
Tsu	Mie ken
Nagoya	Aichi ken
Numazu Hamamatsu	Shizuoka ken
Kofu	Yamanashi ken

Otsu	Shiga ken
Gifu	Gifu ken
Nagano	Nagano ken
Fukushima	Fukushima ken
Ishinomaki	Miyagi ken
Miyako	Iwate ken
Aomori	Aomori ken
Akita	Akita ken
Yamagata	Yamagata ken
Kanazawa	Ishikawa ken
Fushiki	Toyama ken
Fukui	Fukui ken
Matsue	Shimane ken
Sakai	Tottori ken
Okayama	Okayama ken
Hiroshima	Hiroshima ken
Akamagaseki	Yamaguchi ken
Wakayama	Wakayama ken
Tokushima	Tokushima ken
Kochi	Kochi ken
Matsuyama	Ehime ken
Fukuoka	Fukuoka ken
Oita	Oita ken
Saga	Saga ken
Kumamoto	Kumamoto ken
Miyazaki	Miyazaki ken
Kagoshima	Kagoshima ken
Naha	Okinawa ken

NOTIFICATION No. III OF THE DEPARTMENT FOR HOME AFFAIRS.

It is hereby notified that the following changes are made in the positions of Provincial Meteorological Stations mentioned in the Notification No. IV of the 20th year of *Meiji*.

The 12th day, the 2nd Month, the 22nd year of *Meiji*, (1889).

Count MATSUKATA MASAYOSHI
Minister of State for Home Affairs.

Erimo, Soya in Hokkaido are replaced by Kamikawa, Kushiro, Abashiri and Wakkanai in the same.

NOTIFICATION No. XX OF THE DEPARTMENT FOR HOME AFFAIRS.

It is hereby notified that the following changes and additions have been made in the positions of Meteorological Stations mentioned in the Notification No. IV of the 10th month, the 20th year of *Meiji*.

The 10th day, the 6th month, the 24th year of *Meiji*, (1891).

Viscount SHINAGAWA YAJIRO
Minister of State for Home Affairs.

Changes:

Urawa (Saitama ken),	to	Kumagai (the same Prefecture).
Otsu (Shiga ken),	to	Hikone (the same Prefecture).
Matsue (Shimane ken),	to	Hamada (the same Prefecture).

Additions:
Yagi (Nara ken).
Tadotsu (Kagawa ken).

NOTIFICATION No. XLV OF THE DEPARTMENT FOR HOME AFFAIRS.

It is hereby notified that the following change has been made in the position mentioned in the Notification No. IV of the 10th month, the 20th year of *Meiji*.

The 4th day, the 9th month, the 24th year of *Meiji*, (1891).

Viscount SHINAGAWA YAJIRO
Minister of State for Home Affairs.

Mito (Ibaraki ken), to Minato machi (the same Prefecture).

NUMBER OF THE 1ST, 2ND AND 3RD ORDER STATIONS.

There are at present 10 1st order stations in Japan, namely :

Stations.	Lat. N.		Long. E.		Height above mean sea level.	
Wakayama	34°	14'	135°	9'	14.6	m.
Hiroshima	34	23	132	27	4.3	m.
Matsuyama	33	50	132	45	32.4	m.
Tadotsu	34	17	133	46	53.0	m.
Ozaka	34	42	135	31	5.6	m.
Kumamoto	32	48	130	40	16.6	m.
Nagoya	35	10	136	55	15.2	m.
Hakodate	41	46	140	44	3.0	m.
Sapporo	43	4	141	22	16.9	m.
Nemuro	43	20	145	35	26.7	m.

There are 37 2nd order stations, namely :

Stations.	Lat. N.		Long. E.		Height above mean sea level.	
Naha	26°	13'	127°	41'	10.4	m.
Kagoshima	31	35	130	33	3.5	m.
Miyazaki	31	56	131	26	8.0	m.
Kochi	33	33	133	34	42.1	m.
Tokushima	34	06	134	37	3.9	m.
Oita	33	13	131	36	8.5	m.
Yamaguchi	34	11	131	27	32.3	m.
Okayama	34	40	133	54	20.5	m.
Kioto	35	1	135	46	49.4	m.
Saga	33	12	130	18	12.7	m.
Nagasaki	32	44	129	52	57.6	m.
Fukuoka	34	35	130	23	3.8	m.
Itsugahara	34	12	129	16	9.2	m.
Akamagaseki	33	58	130	56	48.2	m.
Sakai	35	33	133	14	2.4	m.
Tsu	34	43	136	28	26.1	m.
Gifu	35	27	136	46	15.0	m.
Hamamatsu	34	43	137	43	27.7	m.
Numazu	35	6	138	51	10.5	m.
Utsunomiya	36	34	139	53	125.0	m.
Choshi	35	44	140	50	2.8	m.
Kanazawa	36	33	136	40	29.0	m.
Fushiki	36	47	137	3	4.3	m.

Stations.	Lat. N.		Long. E.		Height above mean sea level.	
Nagano	36°	40′	138°	10′	420.4	m.
Niigata	37	55	139	3	25.6	m.
Yamagata	38	14	140	17	151.9	m.
Akita	39	42	140	7	10.2	m.
Fukushima	37	45	140	24	62.0	m.
Ishinomaki	38	26	141	19	44.8	m.
Miyako	39	38	141	59	30.4	m.
Aomori	40	51	140	45	4.3	m.
Suttsu	42	48	140	13	16.7	m.
Kamikawa	43	45	142	23	111.0	m.
Soya	45	31	141	55	24.3	m.
Abashiri	44	2	144	14	35.0	m.
Kushiro	43	23	144	28	32.7	m.
Tokachi	42	55	143 '	12	41.0	m.
Erimo	41	55	143	15	63.7	m.

Besides the stations above mentioned we have a great many, which may be regarded as the 3rd order stations. There are 340 stations for the observation of temperature and 232 for that of rainfall. Among them 190 serve for both meteorological elements.

Hokkaido alone has 62 stations for temperature only. Most of them may serve for weather and wind in addition to the elements above mentioned. The contributors are prefectural, municipal, island administration and other public offices as well as from agricultural schools and private individuals.

In this category may be included 55 lighthouse stations which though make adequate observations, for want of sufficient instruments, we have been able to make use of for the wind observations only.

ORGANIZATION OF THE METEOROLOGICAL STATIONS.

Instruments.—The 1st order station is supplied with the following instruments:—

(1) Barometer:

Some stations have been since 1878 using Fortin's cistern barometer made by Casella, London. At other stations, station barometers are used, also made by Casella (closed cistern).

(2) Thermometer:

Thermometers are exposed in louvred screens and the stems are graduated according to Centigrade, but some to Fahrenheit, which when used, is converted to Celsius degrees.

(3) Mason's Dry and Wet Bulb Thermometer.

(4) Maximum (Phillip) and Minimum (Rutherford) Thermometer.

(5) Solar Radiation Thermometer:

This is exposed in the thermometer field.

(6) Terrestrial Radiation Thermometer:

This is exposed on a black board placed in the thermometer field.

(7) Earth Thermometer.

(8) Robinson's Anemometer:

This is erected on a stage over the roof of station.

(9) Anemoscope:

Ditto.

(10) Rainguage:

This is placed on the ground.

(11) Evaporometer:

Ditto.

(12) Jourdan's Sunshine Recorder:

Ditto.

(13) Milne's Seismograph:

This is mounted on a brick stand erected on the ground.

The following instruments are supplied to each 2nd order stations:—

(1) Barometer:

Arranged as above mentioned.

(2) Thermometer:

Ditto.

(3) Mason's Dry and Wet Bulb Thermometer.

(4) Maximum (Phillip) and Minimum (Rutherford) Thermometer.

(5) Robinson's Anemometer:

Arranged as above mentioned.

(6) Anemoscope.

(7) Rainguage:

> As mentioned above.

(8) Milne's Seismograph:

> Ditto.

Barometer and thermometer are all verified at the Central Meteorological Observatory.

Observations at stations, which may be regarded as the 3rd order stations, namely, stations observing rainfall, temperature or other elements only, are generally made once daily, at noon which is chosen as most convenient; as to the 1st and 2nd order stations, the number of observations are fixed by the Notification No. V of 24th year of *Meiji* (1892).

Instructions to the observers.—These instructions have been drawn up according to the decision of the Meteorological Congress and the International Committee, and published at the Tokio Central Observatory. They are furnished to the observers of all stations and are the same as in use at the Tokio Observatory.

Tables for reduction have also been compiled at the Observatory.

Checking of observations.—All the records of observations are scrupulously examined by the staff of the Central Observatory who attend solely to this work, partly by referring to isobars of the daily weather chart. The corrections, calculations and reductions are also examined.

In case any reading is doubtful or erroneous, reference is made at once to the original observations or rough book, by means of letter free of postage to the stations concerned.

However if no fresh light is thrown on the matter and on reconsideration the reading still appears to be wrong, it is rejected.

In such case the monthly average is calculated by omitting it.

Special understanding with the Department for Communications.—All the meteorological records, such as rough books, monthly summaries, 5 days records, yearly summaries &c. are transmitted by post, free of charge, under the special agreement that the Tokio Observatory and all stations shall pay to the post *yen* 50 annually. Telegrams which may be despatched, free of charge, at any time, are as follows:—

(a) Tri-daily regular meteorological telegrams.

(b) Special meteorological telegrams.

(c) Warning meteorological telegrams.

(d) Indication telegrams.

First two are forwarded from the stations to the Tokio Observatory, the other issued from the Observatory to the stations. Every Telegraph Office is instructed from the proper authorities to transmit meteorological telegrams promptly.

SPECIAL OBSERVATIONS.

EARTHQUAKE OBSERVATIONS.

Seismometrical observations were for the first time made at Tokio in the year 1875.
In 1884 the Central Meteorological Observatory requested all parts of the Empire to make earthquake-records and furnished them with blank forms (as shown below). Records have been collected since January 1885. At present the recorders in the districts, cities, towns, etc. number 501, of which 11 attribute to the volunteers. The number of records received at the Central Observatory last year reached over 10,000 and is increasing this year.

THE FORM OF REPORT.

No. Year.		Earthquake Record.
Place	Shocks felt	.
Date of shocks.	Hour, minute, second (Japanese time).	
Duration of shocks.	Minutes and seconds.	
Direction of shocks.	N, NE, E, SE, S, SW, W, NW.	
Intensity of shocks.	Slight, weak, strong, severe.	
Character of shocks.	Motion, horizontal and vertical, and noise if felt.	
Remarks.	Matters not stated above.	
	Name and address of the recorder.	

THUNDERSTORM OBSERVATIONS.

We have a large number of thunderstorm observers; at present they are 143 in number and increasing every year. Last year they contributed 1,331 reports, forms having been distributed from the Observatory.

Instructions to the thunderstorm reporters.—Thunderstorms are to be recorded whenever thunder has been distinctly heard whether accompanied with rain or hail or none.

The Reports are divided into 3 parts, viz, Part A, Part B, and Remarks.

Part A shall contain matters which can be observed without any aid of instrument.

Part B shall contain matters which need some care and instrument such as rainguage; therefore this part is instructed to a reporter possessing a rainguage or to an individual specially voluntcery to report on the subject. Partial reports are accepted also.

In Remarks are to be entered intensity and distance of thunder, aspect, form, direction, and speed of cloud, prospect of thunderstorms, damages, etc. or any other matters which the observer judges to be important, direction in 8 points of compass, rain in 3 degrees—slight, heavy and very heavy—wind force in scales 0–6, and size of hail as compared with any common substance (if it is unusually large, enter its weight also). As the accuracy of time is recommendable, the reporter living in the neighbourhood of a telegraph office or railway station shall keep in his watch or clock the time furnished by either of these establishments.

THE FORM OF REPORT.

No.	Year.	Thunderstorm Record.

PART A.

Prefecture, district, etc.   Date.

Time of thunder distinctly heard ....h......... m.

Its direction...

Rain commenced...h..........m.

Hail commenced... h..........m.

Intensity of rain...

Size of hail ....

PART B.

Direction of thunderstorm passed...................... ....

Last thunder ....h...........m. ....Its direction ....

Rain commenced....h...........m......... ; ceased....h......m ; ......Amount.........

before   during   after

(Thunderstorm.)

Wind direction ....

Wind force ....

Temperature ....

Time of observing temperature...

```
┌─────────────────────────────────────────────────────────┐
│                      REMARKS.                           │
├─────────────────────────────────────────────────────────┤
│                                                         │
│                                                         │
│                                                         │
│                                                         │
│                                                         │
├─────────────────────────────────────────────────────────┤
│                       Reporter's name and address.     │
└─────────────────────────────────────────────────────────┘
```

PHENOLOGICAL OBSERVATIONS.

The system of reporting on periodical phenomena relating to animals and plants has also been organized since some years. Number of places connected with the reports on animals is 24, and on plants 79.

Instructions to the Zoological reporters.—Reports are not to be limitted to the animals specified in the Form; any other animals may be entered in places left blank. Birds are not classified as migratory or not; hence as to migratory birds such as wild-ducks or geese the periods of their arrival and departure are to be specially marked, but as to unmigratory birds as sparrows, crows, etc. their nesting periods are necessary.

As fish and testaceans are not specially distinguished in the Form, the periods of arrival and departure are to be entered for the season fish such as cod fish and herring in Hokkaido and Sphyraena in the southern sea, the periods of spawning for common fish such as dab, etc. In case any fish show an unusual increase their names and rates of increase shall be entered in Remarks. In case any strange bird was seen it shall be described in Remarks. Note singing periods of birds such as cuckoos, nightingales, etc., and also of insects such as cicada etc., and natural hatching periods of silkworms.

THE FORM OF REPORT.

| Zoological Record. | | Date. | | | Prefecture, district, etc. | | | |

Birds.

Names	Date of arrival	Period of nesting	Date of departure	Remarks
Skylark				
Shrike				
Wagtail				
Titmouse				
Nightingale				
Wren				
Wildgoose				
Swallow				
Brown-eared Bulbul				
Woodpecker				
Cuckoo				
Duck				
Quail				
Snipe				
Water rail				
Fish hawk				
Owl				
Crow				
Pheasant				
Sparrow				
&c.				

Fishes and Testaceans.

Names	Date of arrival	Period of spawning	Date of departure	Remarks
Sphyraena				
Salmon				
Herring				
Codfish				
Trout				
Smelt				
Bonit				
Sardine				
Mackerel				
Dab				
Sea-ear				
Oyster				
&c.				

Reptiles.

Names	Period of appearance	Period of disappearance	Remarks
Tortoise			
Lizard			
House-lizard			
Snake			
Viper			
Frog			
Bull frog			
Water lizard			
&c.			

Beasts.

Names	Period of molting	Period of reproduction	Remarks
Sheep			
Ox			
Hare			
Horse			
Deer			
&c.			

Name and address of the reporter.

Instructions to the Botanical reporters.—Reports are not to be limitted to the plants specified in the Form; any other plants may be entered in blank places, mean dates only are to be taken. Dates of sowing are to be specially entered with exception of those plants as mulberry and tea which are not sown yearly. Produce per 1 *Tan* and measure of *Koku* chiefly apply to cereals, but products of either best or worst soil are to be excepted. In case there is any unusual increase or decrease in the products, as in fruitful year or famine, the causes shall be mentioned in Remarks. In case foreign cereals are cultivated the reasons shall be mentioned. As to rice the dates of trans-plantation shall be put in the column for sowing.

THE FORM OF REPORT.

	Species	Date of sowing	Date of budding	Date of appearance of ears	Date of flowering	Date of ripenning or fruit-bearing	Produce of 1 Tan	Koku	Remarks
	Botanical Record.			**Date.**		**Prefecture, district, etc.**			
Corns.	Early rice Ordinary rice Late rice Glutinous rice Upland rice Barley Wheat Rye Oat Germanicum (Millet) Panicum Mi-liaceam Barnyard grass &c.								
Plants of usefull leaves.	Tobacco Indigo Tea Mulberry &c.								
Fruit-bearing Trees.	Grape Pear Orange Chestnut Persimmon Apple &c.								
Flowering Trees.	Plum tree Cherry tree Peach Pyrus specta-bilis Azalea Peony &c.								

Name and address of the reporter.

SNOW DEPTH OBSERVATIONS.

Special snow depth observations are specially made, the stations numbering 71 at present.

Instructions to the snow depth reporters.—The depth of snow shall be recorded in centimetre, but in case of such scale lacking, the Japanese scale may be used. Measure daily at 10ʰ am the depth of snow so long as it covers the ground, record also whether any or no new rain or snowfall has taken place, and the result shall be reported on the 10th, 21st and 31st of each month. Enter in the Remarks, the dates of the commencement and ceasing of snowfall, whether it accompanied rain or hail or none, whether remained on the ground or not, the dates of the first and last rainfall after the ground has been covered with snow; description of heavy snowfall, rapid melting of snow; rapid rising of river and aspect of snow on the neighbouring hills.

THE FORM OF REPORT.

No.	Snow Depth Record.		Date.
	Station.	Prefecture.	District, etc.
Date	Any rain or snow	Depth of snow	Remarks
	Date of despatch	Name of the reporter.	

HIGH LEVEL OBSERVATIONS.

High level meteorological observations have been opened since 1880. First complete series of observations were made on the top of Gozaishodake in the province of Ise near Owari Bay, whose height above the mean sea level is 1,200 metres, during 30 days from September 4th to October 3rd 1888; on the summit of Fujiyama ($3,718^m$) and at Yamanaka (990^m) during 38 days from August 1st to September 7th 1889; on Higashi Hōben (736^m) in the extremity of Western Nippon, during the three months, August, September and October of the same year, and on Ontake ($3,062^m$) and Kuro-sawa (832^m) in Shinano, Central Japan, during 43 days from August 1st to September 12th 1891.

OBSERVATIONS AT SEA.

NOTIFICATION NO. XI OF THE DEPARTMENT FOR HOME AFFAIRS.

It is hereby notified that coasting and foreign-bound vessels specified in the provision of the Art. 6 of the Notification of the Department for Communications No. IV of the 19th year of *Meiji* (1886), shall forward to the Central Meteorological Observatory monthly meteorological observations at sea, according to the following Form, on and after the 1st day of the 1st month of 22nd year of *Meiji*.

The 27th day, the 12th month, the 21st year of *Meiji*, (1888).

Count MATSUKATA MASAYOSHI
Minister of State for Home Affairs.

FORM OF METEOROLOGICAL OBSERVATIONS AT SEA.

Month...year...Meteorological Observations on board the Ship....Captain....from....to....

Date		Position		Wind		Barometer	Thermometer			Cloud				Weather	Sea wave		Sea water		Current		
										Direction											Remarks
Day	Hour	Latitude	Longitude	Direction	Force	Reading	Att. Therm.	Dry	Wet	Upper	Lower	Amount			Direction	Disturbance	Temperature	Density	Direction	Speed	

INSTRUCTIONS TO THE METEOROLOGICAL OBSERVERS ON BOARD SHIPS.

(1) Marine Meteorological Charts are highly useful for navigation and various other purposes and as the reports of meteorological observations at sea furnish important data for the compilation of the Marine Charts, these observations shall be recorded as accurately as possible.

(2) Hours of observations shall be 2ʰam, 6ʰam, 10ʰam, 2ʰpm, 6ʰpm, and 10ʰpm. (Japan time, 135° east of Greenwich) and the observations made at the appointed hours

Shall be without delay recorded in the Report Form. In case any of the observations having been inevitably prevented, the fact shall be shown by drawing a broken line in the Form.

(3) The latitude and longitude of the ship's position at the time of making the observation shall be determined either by an actual measurement or by an estimation, and recorded in the report.

(4) The direction of wind shall be indicated according to 32 points of compass; its force shall be calculated according to the Beaufort scale.

(5) Barometrical pressure shall be in millimetres and temperature in Centigrade degrees; but if the instruments in use are not so graduated, it is desirable to convert the results into millimetres and Centigrade degrees.

(6) As the use of the thermometer attached to a barometer is to know the temperature of mercury in the latter, its readings shall be entered with the barometrical measurements.

(7) Thermometer shall be mounted in a place sufficiently protected from fire or the sun's rays and its indications be entered in the column "Dry."

(8) Readings of wet bulb thermometer are not required, for a vessel is not usually supplied with this instrument, but if supplied before making an observation it is advisable to examine whether the cotton thread of the bulb absorbs sufficient water.

(9) The readings of barometer or thermometer shall be entered without any alteration.

(10) The term "Upper Cloud" means cirrus typed clouds and "Lower Clouds" all other clouds.

(11) In recording cloud amount, clear sky is to be marked 0, and entirely clouded sky 10.

(12) In recording weather the Beaufort notation shall be used.

(13) The heights of waves shall be recorded with minimum or dead calm as 0 and maximum as 9.

(14) As the observations of temperature of sea water are highly useful for purpose of navigation and science, they shall be made as regularly as possible. Pull up a quantity of sea water at the stem of the vessel, immediately plunge a thermometer into and keep it about 15 minutes in the water. Water kept long on board is useless for the purpose. Care must be taken in pulling up the water not to use a bucket warmed by the sun's rays or otherwise for the temperature of the water will be soon affected and it must be remembered also that when the quantity of the water is small, its temperature will be quickly affected by that of the air.

(15) If the density of sea water can be obtained, it shall be recorded also. In case the vessel is provided with a salinometer the degree of the density shall be ascertained and if not, the weight of the water per 1 sho may be entered in the report.

(16) Direction and speed of current shall be entered when they can be ascertained.

(17) In order to make observations on current, a vessel sometimes throws into the sea bottles each containing a description of her position, namely, the latitude, longitude, and date and hour. In case therefore any vessel finds a bottle floating on sea, it is advisable to pick it up, examine its contents and take down the date etc., given in the description in the column for "Remarks," but care must be taken not to throw it again into the sea without adding to it the vessel's position, date etc. In case there are useless bottles on board, they shall be used in this way; good data will be thus obtained for investigating our sea currents.

(18) During a storm frequent observation shall be made besides regular hours, entering the results in the Remark column.

(19) The meteorological instruments, especially barometer and thermometer shall be well selected; in case they are not true in their indications, great mistakes will be caused. If any person desires to have his instruments tested, he shall apply to the Central Meteorological Observatory or to provincial meteorological stations.

(20) Index errors of barometer and thermometer shall be entered in the annexed Form and be sent in together with the Report.

Barometer.		Thermometer.					
		Thermometer attached to Barometer		Dry		Wet	
Readings	Errors	Readings	Errors	Readings	Errors	Readings	Errors

Reports received from various vessels at the Central Observatory in 1891 are as follows:—

	Number of vessels.	Reports.
Japanese men-of-war	10	64
Vessels belonging to Department for Communications	1	—
Merchant ships	54	963
	65	1,027
Foreign men-of-war	1	3
Foreign merchant ships	11	85
	12	88
Total	77	1,115

TELEGRAPHIC WEATHER SERVICE.

The service of the weather telegraphy has been organized in February 1883.

The principal operations connected with the preparation and issue of the Daily Weather Report and Storm Warnings are as follows:—

The service receives, when the telegraphic communications are perfect, 45 cypher reports tri-daily, that is, 6^h am, 2^h pm and 10^h pm Japan time.

The foreign reporting stations 5 in number extend along the part of the eastern coast of the Asiatic Continent, namely Manila, Hongkong, Amoy, Shanghai and Vladivostok.

An information is received twice a day in accordance with an arrangement agreed between the Central Observatory, the Great Northern Telegraph Company and the Eastern Extension Telegraph Company. These informations are transmitted through the courtesy of these companies by their private lines, free of charge. In return the Central Observatory sends its reports of meteorological observations taken at 10^h am and 2^h pm daily.

The majority of morning telegrams arrive in Tokio at about 7^h am, those of afternoon and night at about 3^h and 11^h pm respectively. As fast as the reports come in, the information is entered in the working chart, which shows for each station the barometrical and thermometrical readings with their respective alterations during the preceding 8 and 24 hours, and the direction and force of wind and the state of weather, together with any changes of importance which may have been noticed in the course of the preceding hour. From this chart, which is preserved in the office, another chart is drawn for publication, as described further on.

If necessary, telegraphic intelligence of a storm or of atmospheric disturbance is immediately sent to the coasts and interior of the Empire.

Since June 1st, 1884, the forecast of weather for the following day has been prepared in every morning, but not for publication. By this experience, however, the staff has been much trained in this business. Since 1888 Indications have been published every evening at 9^h pm and since June 1st, 1891, at 2^h pm. Thus the forecasts for 7 districts in Japan are drawn up and printed in weathermaps which are forwarded to all meteorological stations in the Empire, every official department, certain officers, companies, private individuals and many of the Tokio newspapers. Most provincial newspapers receive information from the provincial stations. In addition to the regular issuance of forecasts, the service sometimes answers by telegram the inquiries as to probable weather, for not more than one day in advance. The districts for which the forecasts are prepared are as follows:—

1	South.	5	Northwest.
2	Inland sea.	6	East.
3	West.	7	North.
4	Southeast.		

The daily weathermap which is issued once a day with tri-daily reports in one sheet, is drawn on transfer paper and is ready at about 4ʰ pm, when it is at once sent to a lithographer to be printed. The copies for delivery by hand are issued by the lithographer at about 4ʰ 30ᵐ pm while the remaining is transmitted by post to the provincial stations, subscribers and others. Weathermaps to be exhibited at the Ueno and Shimbashi Railway Terminus, and report to be printed in Official Gazette are sent out as soon as possible. Weathermaps contain a map of Japan and Eastern Asia, observations made at 45 stations with a summary of the weather of Japan, the isobaric and isothermal lines printed in black on blue ground, a forecast for the next day ending at 6ʰ pm, and some China coast and Siberian reports.

A constant watch is kept during a day and night, and whenever the telegrams come in at about 7ʰ am, 3ʰ pm and 11ʰ pm, and also on the receipt of any special telegram, the condition of the weather is carefully considered; if it appears threatening, storm warnings at once dispatched to such parts of the Empire as are thought to be menaced by a gale.

VERIFICATIONS OF INDICATIONS AND STORM WARNINGS.

The average successes of the weather forecast and storm warnings during 10 years are shown in the following:—

PERCENTAGE OF MEAN MONTHLY VERIFICATIONS

Indications of

Months.	Weather.	Wind direction.	Storm warnings.
January	85	80	74
February	78	82	72
March	81	81	78
April	76	74	69
May	79	82	69
June	81	81	68
July	83	79	58
August	84	84	65
September	82	79	63
October	80	84	67
November	79	74	77
December	85	84	76

PERCENTAGE OF YEARLY VERIFICATIONS.

Indications of

Year.	Weather.	Wind direction.	Storm warnings.
1883	72	73	71
1884	85	84	58
1885	83	86	51
1886	88	91	60
1887	83	83	68
1888	80	79	76
1889	78	77	84
1890	77	81	76
1891	80	82	75
1892	83	80	69
Mean	81	82	69

NOTIFICATION OF CENTRAL METEOROLOGICAL OBSERVATORY AS TO TERMS USED IN THE WEATHER INDICATIONS.

It is hereby notified that terms used in the Weather Indications have been revised and the following terms shall be used on and after the 15th day, the 6th month, the 25th year of *Meiji*.

The 6th month of the 25th year of *Meiji*, (1892).

CENTRAL METEOROLOGICAL OBSERVATORY.

TERMS FOR THE WEATHER INDICATIONS.

Wind.	Weather.
Northerly wind	Fair
Northerly or Easterly wind	Changeable
Easterly wind	Cloudy
Southerly or Easterly wind	Rain
Southerly wind	Snow
Southerly or Westerly wind	Clearing later
Westerly wind	Fog
Northerly or Westerly wind	Colder
Variable wind	Warmer

INSTRUCTIONS REGARDING PROVINCIAL WEATHER FORECAST.

(1) According to the provision of Art. 2 of Notification of the Department for Home Affairs No. V of the present year (1892) the weather forecast to be issued from a provincial meteorological station is confined to the following 3 items;

(a) Weather.

(b) Direction of wind.

(c) Temperature.

(2) Provincial weather forecast shall be drawn up with the terms determined by the Central Observatory.

(3) Provincial weather forecast shall be issued at 6ʰ pm every day and indicate the weather for the following 24 hours.

(4) District to be included in a provincial weather forecast shall be determined on the approval of the Central Observatory.

(5) Verification of the provincial weather forecast shall be made in accordance with the method determined by the Central Observatory.

(6) Verification of the provincial weather forecast shall be forwarded to the Central Observatory before the 10th of the following month.

CAUTIONARY STORM SIGNALS.

If strong winds or gales are deemed probable at or near any of the undermentioned signalstations, cautionary signals will be hoisted at the respective stations.

SIGNALS.

● A red ball hoisted during the day means: "Strong winds are probable."

● Corresponding nightsignal: one red light.

▲ A red cone, point upwards hoisted during the day means: "Very serious disturbance are probable."

●● Corresponding nightsignal: two red light in a horizontal position.

In cases of approaching atmospheric disturbance meteorological telegrams will be posted up at the signal stations and other public places. Interested parties are advised to consult the same, whether a signal is ordered in them and hoisted or not.

Orders for a signal hold good for 24 hours from the time of issue.

SIGNALSTATIONS.

NOTE:—Stations without remark are fully equipped as above.
The partial equipment of a station is indicated under Remark by one or more of the letters B (Ball); C (Cone); L (Red light).
Thus: B C under Remark denotes that the corresponding station has Ball and Cone, or daysignals only, but no nightsignals.
For Districts, see Map.

Dis-trict	Num-ber	Name of Station	Location	Remark	Established on the	by
I	1	Kagoshima	Kiushu South		1 Dec. 1883	Kagoshima Ken
I	2	Susaki	Shikoku South	B C	1 May 1888	Kochi Ken
I	3	Kochi	Shikoku South	B C	1 May 1888	Kochi Ken
I	4	Wakayama	Kii Channel		15 Sept. 1884	Wakayama Ken
I	5	Osaki	Kii Channel	B L	1 April 1892	Wakayama Ken
I	6	Yuasa	Kii Channel	B L	1 April 1892	Wakayama Ken
I	7	Yura	Kii Channel	B L	1 April 1892	Wakayama Ken
I	8	Gobo	Kii Channel	B L	1 April 1892	Wakayama Ken
I	9	Tanabe	Kii Channel	B L	1 April 1892	Wakayama Ken
II	10	Saiki	Bungo Channel	B L	15 June 1892	Oita Ken
II	11	Saganoseki	Bungo Channel	B	20 April 1891	Oita Ken
II	12	Oita	Bungo Channel		1 Feb. 1890	Oita Ken
II	13	Nagasu	Kiushu North	B	1 Oct. 1891	Oita Ken
II	14	Nakatsu	Kiushu North		1 Feb. 1890	Oita Ken
II	15	Hiroshima	Inland Sea		25 Nov. 1887	Hiroshima Ken
II	16	Kure	Inland Sea		1 Oct. 1889	Admiralty Station
II	17	Onomichi	Inland Sea		10 Aug. 1884	Hiroshima Ken
II	18	Mitsugahama	Shikoku North		15 June 1886	Ehime Ken
II	19	Takamatsu	Shikoku North	B L	20 June 1888	Kagawa Ken
II	20	Kobe	Inland Sea		15 Sept. 1887	Hiogo Ken
II	21	Temposan	Inland Sea	B L	1 Aug. 1884	Ozaka Fu
II	22	Osaka	Ozaka bay		1 April 1890	Ozaka Fu
II	23	Moto-Torishima	Ozaka bay	B L	1 Nov. 1891	Ozaka Fu
II	24	Kizugawa	Inland Sea	B L	15 June 1892	Ozaka Fu
II	25	Sakai	Ozaka bay	B L	1 Nov. 1885	Ozaka Fu
III	26	Misumiminato	Kiushu West		25 Dec. 1887	Kumamoto Ken
III	27	Saga	Shimabara gulf	B L	1 Dec. 1890	Saga Ken
III	28	Karatsu	Kiushu North	B L	1 Dec. 1890	Karatsu Sekitan Kaisha
III	29	Nagasaki	Kiushu West		15 Oct. 1884	Nagasaki Ken
III	30	Sascho	Kiushu Northwest		29 March 1890	Admiralty Station
III	31	Itsugahara	Tsushima	B	1 April 1887	Nagasaki Ken
III	32	Miike	Shimabara gulf	B L	1 Jan. 1886	Fukuoka Ken
III	33	Wakatsu	Shimabara gulf		1 Feb. 1890	Fukuoka Ken
III	34	Hakata	Kiushu North		10 Sept. 1885	Fukuoka Ken
III	35	Wakamatsu	Kiushu North	B L	15 Aug. 1887	Fukuoka Ken
III	36	Akamagaseki	West coast		1 Jan. 1886	Yamaguchi Ken
IV	37	Kushimoto	SE^{rn} Kii	B L	1 April 1892	Wakayama Ken
IV	38	Miwazaki	SE^{rn} Kii	B L	1 April 1892	Wakayama Ken
IV	39	Shingu	SE^{rn} Kii	B L	1 April 1892	Wakayama Ken
IV	40	Yokkaichi	Owari bay		15 Sept. 1888	Miye Ken
IV	41	Atsuta	Owari bay		10 Oct. 1888	Aichi Ken
IV	42	Nagoya	Owari bay	B L	1 July 1890	Aichi Ken
IV	43	Gifu	North of Nagoya		1 Jan. 1884	Gifu Ken
IV	44	Kaketsuka	Near Hamamatsu		15 Jan. 1886	Shizuoka Ken
IV	45	Shimizu	Suruga gulf		20 Dec. 1884	Shizuoka Ken
IV	46	Yokosuka	Tokio bay		1 Sept. 1880	Admiralty Station
IV	47	Shinagawa	Tokio bay		1 Dec. 1883	Centr. Met'l. Observatory
IV	48	Choshi	Near Inuboye		20 July 1889	Chiba Ken
V	49	Mikawa	Northwest coast		1 May 1892	Ishikawa Ken
V	50	Ataka	Northwest coast		1 Oct. 1892	Ishikawa Ken
V	51	Kanazawa	Northwest coast		11 April 1885	Ishikawa Ken

Dis-trict	Number	Name of Station	Location	Remark	Established on the	by
V	52	Kauui-Kanaiwa	Northwest coast	B L	11 April 1885	Ishikawa Ken
V	53	Shimo-Kanaiwa	Northwest coast		15 Sept. 1891	Ishikawa Ken
V	54	Nanao	Noto		11 April 1885	Ishikawa Ken
V	55	Fushiki	Northwest coast		15 April 1886	Toyama Ken
V	56	Hignshi-Iwaso	Northwest coast		15 Oct. 1887	Toyama Ken
V	57	Niigata	Northwest coast		10 May 1885	Niigata Ken
V	58	Kamo	Northwest coast		15 July 1891	Yamagata Ken
V	59	Sakata	Northwest coast		10 Jan. 1885	Yamagata Ken
V	60	Tsuchizaki	Northwest coast	B L	1 Oct. 1892	Akita Ken
V	61	Noshiro	Northwest coast		15 Jan. 1889	Akita Ken
VI	62	Ishinomaki	Sendai bay		1 Sept. 1885	Miyagi Ken
VII	63	Aomori	Tsugaru strait		1 Dec. 1883	Aomori Ken
VII	64	Fukuyama	Tsugaru strait	B C	1 Jan. 1884	Hokkaido Cho
VII	65	Hakodate East	Tsugaru strait		1 Jun. 1884	Hokkaido Cho
VII	66	Hakodate West	Tsugaru strait		1 Jan. 1884	Hokkaido Cho
VII	67	Mori	Volcano bay	B C	25 Jan. 1885	Hokkaido Cho
VII	68	Esashi	Hokkaido West	B C	10 Dec. 1884	Hokkaido Cho
VII	69	Mororan	Volcano bay		15 July 1884	Hokkaido Cho
VII	70	Suttsu	Hokkaido West		1 Jan. 1884	Hokkaido Cho
VII	71	Iwanai	Hokkaido West		1 July 1884	Hokkaido Cho
VII	72	Shikonai	Hokkaido West		10 May 1884	Hokkaido Cho
VII	73	Mashike	Hokkaido West	B C	1 July 1892	Hokkaido Cho
VII	74	Nemuro	Hokkaido East		5 Oct. 1887	Hokkaido Cho
VII	75	Kushiro	Hokkaido East		1 April 1890	Hokkaido Cho

RULES RELATING TO METEOROLOGICAL TELEGRAMS.

SECTION I.

INSTRUCTION.

§ 1.—Meteorological telegrams are those telegrams which are exchanged between the Central Meteorological Observatory, Provincial Meteorological Stations and Storm Signal Stations.

§ 2.—Meteorological telegrams are divided into 6 kinds

(a) Regular telegrams.

.(b) Extraordinary telegrams.

(c) Warning telegrams.

(d) Indication telegrams.

(e) Weather telegrams.

(f) Service telegrams.

§ 3.—Regular and extraordinary telegrams shall be written out with Japanese figures only, and telegrams of other kinds with Katakana (Japanese Alphabet) only, with exception of extraordinary telegram from the Central Observatory and service telegram, having certain particular forms.

SECTION II.

REGULAR TELEGRAMS.

§ 4.—Regular telegrams are those telegrams which are sent from every reporting station to the Central Observatory at 6ʰ am, 2ʰ pm and 10ʰ pm daily.

§ 5.—In the despatch form of a regular telegram shall be marked the words "Regular meteorological telegram" in red ink, and the name of the station shall be omitted and the word "Meteorological" be written in place of "Central Meteorological Observatory."

§ 6.—The telegram shall be written out with 20 figures according to the following order:

§ 7.—In recording atmospheric pressure, it shall be reduced to its equivalent at sea level and 45° latitude, rejecting the figure 7 from 700 and carrying the figures down to tenths of a millimetre.

§ 8.—Wind directions shall be given in accordance with 8 points of the compass, using the following marks:

Wind directions.	Marks.
Calm.	0
N E	1
E	2
SE	3
S	4
SW	5
W	6
N W	7
N	8

§ 9.—Wind force shall be calculated from its velocity at 20 minutes before the time of observation, the scale being from 0 to 6.

§ 10.—Precipitation amount shall be given down to millimetres taking the amount for last eight hours. In case the amount of melted snow can not be obtained, the depth of snow shall be given down to centimetres.

§ 11.—Cloud directions shall be recorded in accordance with 8 points of the compass, using the marks shown in § 8.

§ 12.—Speed of clouds shall be given with the following marks:

Speed.		Marks.
Stationary or uncertain		0
Upper cloud	slow.	1
	fast.	2
	rapid.	3
Middle cloud	slow.	4
	fast.	5
	rapid.	6
Lower cloud	slow.	7
	fast.	8
	rapid.	9

§ 13.—Temperature shall be recorded down to tenths of a degree.

§ 14.—Cloud amount shall be recorded with the following marks:

Amount.	Marks.
0	0
1	1
2	2
3	3
4	4
5	5
6	6
7	7
8	8
9,10	9

§ 15.—Forms of clouds shall be recorded with the following marks:

Forms.		Marks.
Upper cloud	cirrus.	0
	cirro-stratus.	1
Middle cloud	cirro-cumulus.	2
	cumulo-cirrus or alto cumulus.	3
	strato-cirrus or alto stratus.	4
Lower cloud	strato-cumulus.	5
	nimbus.	6
	cumulus.	7
	cumulo-nimbus.	8
	stratus.	9

§ 16.—Weather shall be recorded with the following marks:

Weather.	Marks.
No remarks.	0
Rain.	1

Weather.	Marks.
Snow.	2
Shower.	3
Fog.	4
Graupel.	5
Hail.	6
Thunderstorm.	7
Thunder.	8
Frost.	9

§ 17.—Remarks shall be given with the following marks:

Remarks.	Marks.
No remarks.	0
Barometer falling (2—4 mm in 4 hours).	1
Barometer falling rapidly (above 4 mm in 4 hours).	2
Barometer lowest.	3
Barometer rising rapidly (above 4 mm in 4 hours).	4
Threatening weather.	5
Amount of melted snow.	6
Depth of snow.	7
Scud.	8
Rolling sea.	9

§ 18.—Minimum temperature shall be reported at 2^h pm, Maximum at 10^h pm.

SECTION III.

EXTRAORDINARY TELEGRAMS.

§ 19.—Extraordinary telegrams include those telegrams which are sent from the Central Observatory to the Provincial Stations requiring them to make special observations, telegrams in reply to above and telegrams from the stations to the Observatory informing unusual meteorological phenomena.

§ 20.—An extraordinary telegram shall be marked on its despatch form the words "Extraordinary Meteorological Telegram" in red ink, and the name of the station shall be omitted, and when addressed to the Central Meteorological Observatory the words "Meteorological" only be marked.

§ 21.—An extraordinary telegram from the Observatory shall consist of 3 *Katakana* representing the nature of information required.

§ 22.—In the extraordinary telegrams sent from the Provincial Stations to the Observatory, meteorological phenomena are represented by 11 figures in the following order:

Air pressure. Wind direction. Wind force. Cloud direction. Cloud speed. Weather. Remarks. Time.

§ 23.—Figures representing air pressure, wind direction, wind force, cloud direction, cloud speed, weather and remarks shall be the same as those used in regular telegrams.

§ 24.—Hours shall be represented, e. g.

1^h am by 01, 1^h pm by 13, 2^h pm by 14, etc.

§ 25.—Whenever the Observatory requires to be informed of the minimum air pressure, maximum wind velocity etc., the stations shall reply in accordance with the example (b) given § 48, of these Rules.

§ 26.—In extraordinary telegrams from the Stations to the Observatory, seismological phenomena shall be represented by 9 figures in the following order:—

Intensity. Hour. Minutes. Seconds. Remarks.

§ 27.—The intensity of earthquake shall be represented by the following marks:

Intensity.	Marks.
Slight shock.	01
Feeble shock.	02
Violent shock.	03
Most violent shock.	04

§ 28.—Hours shall be represented in the same manner as § 24.

§ 29.—Remarks on earthquake shall be represented by the following marks:

Remarks.	Marks.
No damage.	0
Clock stopped, articles on shelves fallen down.	1
Walls cracked or fallen down.	2
Houses destroyed.	3
Many houses destroyed.	4
Ground cracked.	5
Subterranian noise.	6
Waves innundating.	7
Shocks following successively.	8

SECTION IV.

WARNING TELEGRAM.

§ 30.—Warning telegrams are those telegrams which are forwarded from the Observatory to certain Provincial Stations and other places to warn them of bad weather to be expected in the districts concerned.

§ 31.—In case the coast of any district is warned, the warning of the same purport shall be despatched to the interior of the same district for the purpose of precaution.

§ 32.—In case it is deemed necessary to warn the interior after the coast has been warned, the warning telegram shall be despatched anew to the interior.

§ 33.—The purport of a warning telegram shall be represented by 10 *Katakana* in the following order :—

§ 34.—In case the position of the lowest pressure and its progressive motion are uncertain, or when specially required, the warning telegram is written out in the following order :—

§ 35.—In case the condition of the disturbance is to be further informed after a warning has been issued, the telegram shall be written out with 6 *Katakana* in the following order :—

§ 36.—Cyphers to be used in warning telegrams are shown in the Appendix.

SECTION V.

INDICATION TELEGRAM.

§ 37.—Indication telegrams are those telegrams which are despatched from the Observatory to the stations and other places, containing the weather forecast of the Empire.

§ 38.—An indication telegram shall be drawn up in the following order :—

(a) Weather forecast of each district—arranged in the order of districts.
(b) Position of highest pressure.
(c) Reading of highest pressure.
(d) Position of lowest pressure.
(e) Reading of lowest pressure.

§ 39.—Cyphers used for the above items are contained in the Appendix.

§ 40.—In case an indication of one or more districts is same as the preceding or no indication is issued thereof, the blank shall be filled up by a cypher shown in the Table IV of the Appendix.

§ 41.—In case the position of high or low pressure is not well defined, indication telegram consists of the forecast only.

SECTION VI.

WEATHER TELEGRAM.

§ 42.—Weather telegrams are those telegrams which are despatched from the Observatory in compliance with the requests of the Provincial Stations, containing the general aspect of weather in the Empire.

§ 43.—The weather telegram gives air pressure, wind direction and weather of 15 stations in the following order :—

1	Kagoshima	6	Sakai	11	Ishinomaki
2	Kochi	7	Hamamatsu	12	Hakodate
3	Ozaka	8	Tokio	13	Sapporo
4	Nagasaki	9	Kanazawa	14	Nemuro
5	Akamagaseki	10	Akita	15	Fusan

§ 44.—The weather telegram shall be drawn up with 60 *Katakana* in the following order :—

In case no telegram is received from any station, the blank shall be filled up by a cypher.

Kagoshima.	Sakai.	Ishinomaki.
0 0 } Air pressure	0 0 } Air pressure	0 0 } Air pressure
0 Wind direction	0 Wind direction	0 Wind direction
0 Weather	0 Weather	0 Weather
Kochi.	Hamamatsu.	Hakodate.
0 0 } do	0 0 } do	0 0 } do
0	0	0
0	0	0
Osaka.	Tokio.	Sapporo.
0 0 } do	0 0 } do	0 0 } do
0	0	0
0	0	0
Nagasaki.	Kanazawa.	Nemuro.
0 0 } do	0 0 } do	0 0 } do
0	0	0
0	0	0
Akamagaseki.	Akita.	Fusan.
0 0 } do	0 0 } do	0 0 } do
0	0	0
0	0	0

§ 45.—Cyphers to be used in the weather telegrams are shown in the Appendix.

SECTION VII.

SERVICE TELEGRAM.

§ 46.—Service telegrams are those telegrams which pass between the Central Observatory and Provincial Stations or between Stations.

§ 47.—Cyphers are shown in the Appendix.

§ 48.—When an application is made for a meteorological observation, telegram in reply shall drawn up in accordance with one of the following forms:—

(a) 17 Figures.

0 0 } 0 Air pressure 0 Wind direction 0 Wind force 0 0 } 0 Precipitation 0 Cloud direction 0 Cloud speed 0 0 } 0 Air temperature 0 Cloud amount 0 Cloud form 0 Weather 0 Remarks

(*b*)   10 Figures.

Lowest pressure · Time · **Wind direction** · **Max. wind force** · Time

O͞O|O   O|O   O   O|O   O|O

(*c*)   8 Figures (rejecting decimals in pressure and precipitation).

Air pressure · Precipitation · Hour of duration · O Remarks

O|O   O|O|O   O|O   O

(*d*)   6 Figures (rejecting decimals in pressure).

Air pressure · O Wind direction · O Wind force · O Weather · O Remarks

O|O   O   O   O   O

§ 49.—Cyphers to be used in the above forms are the same as in regular telegrams.

Appendix will not be produced here.

RULES RESPECTING METEOROLOGICAL SIGNALS.

The following Rules respecting to Meteorological Signals are hereby established and shall be put in force on and after the 15th, the 6th month, the 25th year of *Meiji*.

The 6th month, the 25th year of *Meiji*, (1892).

CENTRAL METEOROLOGICAL OBSERVATORY.

RULES RESPECTING METEOROLOGICAL SIGNALS.

(1)   Meteorological signals are of 2 varieties; namely,

    (*a*)   Storm signals

    (*b*)   Weather signals

(2) The signals shall be displayed, whenever the storm warnings are received from the Central Observatory, for 24 hours from the time of despatch.

The whole text of a warning telegram shall be placarded beneath the signals.

(3) Storm signals are divided into 2 classes; viz.,

Red Ball and Red Cone.

In night a red lamp is used in place of the Ball and two lamps abreast in place of the Cone.

Red Ball

Red Lamp

Red Cone

Two Red Lamps

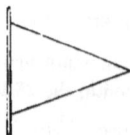

(4) Red Ball means "strong wind expected," Red Cone (pointed upward) "Very serious disturbance expected".

(5) Weather signals shall be displayed when the weather indications are received from the Central Observatory or Provincial Stations.

(6) Weather signals are divided into 3 classes; viz.,

    (a) Triangular Flag

    (b) Square Flag

    (c) Streamer

(7) Triangular Flag indicates wind direction and it is classified as follows:—

Northerly wind.

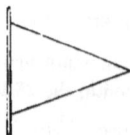

Easterly wind.

Southerly wind.

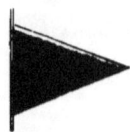

Westerly wind.

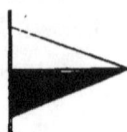

(8) Square Flag indicates weather and is classified as follows :—

Fair.	Cloudy.	Rain or Snow.	Changeable.

Snow.	Fog.	Frost.

(9) Streamer indicates the rise and fall of temperature and is classified as follows :—

Colder.	Warmer.

CENTRAL METEOROLOGICAL OBSERVATORY.

STAFF ORGANIZATION.

The Regulations relating to the Staff Organization of the Central Meteorological Observatory have been established by an Imperial Ordinance.

We hereby give Our Sanction to the Staff Organization of the Central Meteorological Observatory and order the same to be promulgated.

(H. I. M.'s Sign Manual)
(Privy Seal)

The 2nd day, the 8th month, the 23rd year of *Meiji*, (1890).

Count YAMAGATA ARITOMO
Minister President of State.

Count SAIGO YORIMICHI
Minister of State for Home Affairs.

IMPERIAL ORDINANCE No. CLVI.—STAFF ORGANIZATION OF THE CENTRAL METEOROLOGICAL OBSERVATORY.

ART. 1.—The Central Meteorological Observatory belongs to a control of the Home Minister and deals with the following matters:—

    (a) Meteorological observations
    (b) Meteorological reports
    (c) Meteorological investigations
    (d) Verifications of meteorological instruments
    (e) Weather forecast
    (f) Storm warning
    (g) Seismological observations
    (h) Observations of terrestrial magnetism
    (i) Observations of atmospheric electricity
    (j) Air analysis

ART. 2.—The staff is composed of:—

    (a) Director
    (b) Titular meteorologist
    (c) Titular meteorologist Probationer
    (d) Assistant meteorologist
    (e) Clerk

Art. 3.—The director shall be appointed from the titular meteorologists, and he shall superintend, under the supervision of the Minister for Home Affairs, the operations of the staff.

Art. 4.—Titular meteorologist shall be of *so-nin* rank and four in number; they shall perform duties allowed to them by the order of the director. Number of probationer shall be one.

Art. 5.—Assistant meteorologist shall be of *han-nin* rank and fifteen in number; they shall perform their duties under the order of superior officers.

Art. 6.—Clerks who shall be of *han-nin* rank and five in number, shall perform miscellaneous duties under the order of superior officers.

Art. 7.—Division of the business shall be determined by the Minister for Home Affairs.

AMENDMENT IN THE STAFF ORGANIZATION.

We hereby give Our Sanction to the present amendment in the Staff Organization of the Central Meteorological Observatory and order the same to be promulgated.

(H. I. M.'s Sign Manual)
(Privy Seal)

The 24th day, the 7th month, the 24th year of *Meiji*, (1891).

Count MATSUKATA MASAYOSHI
Minister President of State.

Viscount SHINAGAWA YAJIRO
Minister of State for Home Affairs.

IMPERIAL ORDINANCE No. CVI.

The following amendment has been made in the Organization of the Central Meteorological Observatory.

In the provision of ART. 4 "four in number" shall be amended "three in number."

In the provision of ART. 5 "fifteen in number" shall be amended "ten in number."

In the provision of ART. 6 "five in number" shall be amended "three in number"

ADDITIONAL RULES.

The above Ordinance shall take effect on the 16th day, the 8th month, the 24th year of *Meiji*.

OPERATIONS OF THE CENTRAL METEOROLOGICAL OBSERVATORY.

The functions of the Observatory are divided into three sections, namely :—

(a) Service of Observations.
(b) Service of Statistics.
(c) Service of Indications.

The business transacted in the Service of Observations consist of :—

(a) Matters relating to meteorological observations.
(b) Matters relating to seismic observations.
(c) Matters relating to terrestrial magnetism.
(d) Matters relating to atmospheric electricity.
(e) Matters relating to air analysis.
(f) Matters relating to the verification of meteorological instruments.

In the Service of Statistics :—

(a) Matters relating to meteorological reports.
(b) Matters relating to meteorological researches.

In the Service of Indications :—

(a) Matters relating to weather forecasts.
(b) Matters relating to storm warnings.

PRINCIPAL INSTRUMENTS.

The following are the principal instruments used at the Observatory :—

Rédier's Thermograph.
Richard Thermograph.
Kew Hygrograph.
Fuess Hygrometer.
Mason's Dry and Wet Bulb Thermometers.
Fuess Maximum Thermometer.
Fuess Minimum Thermometer.
Casella's Maximum Solar Radiation Thermometer.
Casella's Minimum Terrestrial Radiation Thermometer.
Casella's Maximum Solar Radiation Thermometer in Vacuo.
Conjugate thermometers.
Earth Surface Thermometer.
Earth Thermometer—7 m. 5 m. 3 m. 1.2 m. 0.6 m. & 0.3 m.
Rainguage (receiver 2 decimetre in diameter).
Evaporometer   (ditto).
Robinson's Anemograph and Electro-Anemometer.
Jourdan's Sunshine Recorder.

Kampbel's Sunshine Recorder.
Robinson's Standard Anemometer.
Fortin's Mercurial Barometer.
Rédier's Barograph.
Richard Barograph.
King's Barograph.
Fuess Syphon-Fortin Mercurial Barometer.
Fuess Standard Thermometer.
Vacuometer.
Cathetometer.
Comparing Apparatus of Thermometers.
Meterglass-verifying Instruments.
Air-pump for comparing Aneroid.
Magnetometer.
Mascart's Magnetgraph.
Mascart's Electrograph.
Milne-Gray's Seismograph.
Portable Seismometer.

PERSONNEL AND THEIR APPOINTMENTS.

The director receives a yearly salary of *yen* 1,400, a titular meteorologist *yen* 1,000 and a probationer *yen* 600.

The Chief of Service is appointed from the titular meteorologists. Assistant meteorologist receives a monthly pay varying from *yen* 40 to 12.

Official residences are provided in the enclosure of the Observatory for the Chief of Service of Indications and assistant meteorologists who are engaged in the telegraphic matters.

A clerk receives a monthly pay varying from *yen* 15 to 12.

The salaries are all payed by the Government, and generally the promotion takes place after a continued service of 3 years as to the officers of *so-nin* rank, 1 year as to those of *han-nin* rank. All the officers have claim to superannuation after 60 years of age and 15 years of service, the life pension being equal to one-fourth of their salaries at the time of retirement.

The titular meteorologists are nominated by the Minister President of State with the approval of the Emperor and the other officials by the Minister for Home affairs.

Besides the officers above mentioned, many subordinates with monthly salaries below *yen* 12 are employed by the director. A librarian is appointed from the subordinate employés.

The Chief of a Provincial Station receives a monthly salary of about *yen* 80 to 15, and the assistant a monthly salary below *yen* 25. Their salaries are borne by the Revenue, and they are appointed by the Governor of Prefecture concerned.

PUBLICATIONS OF THE CENTRAL METEOROLOGICAL OBSERVATORY.

Publications of the Central Meteorological Observatory since its opening have been as follows :—

(1) Report of the daily meteorological observations of Tokio and other stations.

(2) Report of the meteorological observations for each 5 day periods made at Tokio.

(3) Comparison of the meteorological observations made at several meteorological stations in Japan.

(4) Results of the meteorological observations made at Tokio, for the lustrum, 1876-1880.

(5) Report of the meteorological observations made at Tokio for the 10 years, viz. 1876-1885.

(6) Monthly and yearly means, extremes and sums of the meteorological observations for the provincial stations.

(7) Report of the meteorological observations in the Empire of Japan.

(8) Annual meteorological report for the year 1886, consisting of Part I and II (Part II containing the results of hourly and miscellaneous observations in Tokio Central Observatory). 1887 Part I, II, 1888 Part I, II. (I, 31 stations ; II, Tokio hourly observations and simultaneous observations on the Gozaishodake and Yokkaichi). 1889 Part I, II. (I, 34 stations ; II, Tokio hourly observations and simultaneous observations on the Fujiyama and Yamanaka). 1890 Part I, II (I, 44 stations ; II Tokio hourly observations and memoir on the low pressure in Japan).

(9) Tri-daily weathermaps, reports, remarks and indications from March 1883.

(10) Monthly summaries and monthly means with maps (areas of high and low pressure, monthly isobars, isotherms, rainfall ; storm warnings, indications, etc.) from 1883.

(11) Tokio meteorological report for the year 1885, prepared from three hourly observations.

(12) Hourly meteorological observations for the year 1890 (Tokio.)

(13) Instruction for meteorological observers (Japanese text 1880).

(14) James Glaisher's hygrometrical tables (Japanese translation 1880). .

(15) The typhoon of September 13, 14, 1881 (Japanese text).

(16) The typhoon of September 26, 27, 1881 (Japanese text).

(17) Forms of clouds (colored sheets).

(18) Additional instructions to meteorological observers (lithographed in English and Japanese, 1882).

(19) Comparison of scales (Graphical sheet, 1882).

(20) Cautionary signals and signal stations (Annual publication).

(21) Report of the seismometrical observations for 1885-1889 (Japanese text).

(22) Some researches on agricultural meteorology Part I-V (Japanese text).

(23) Instruction for meteorological observers with tables (Japanese text 1886).
(24) Typhoon tracks near Japan, with hints for seamen (Translated from the German).
(25) Approximate normal pressure, temperature, and rainfall in Japan (1887).
(26) Report of an expedition to Mount Fuji 1887, 1888.
(27) Meteorological observations made at Yokkaichi and on the top of Mount Gozaishodake during the typhoon season in the year 1888 (Japanese text).
(28) Organization du service météorologique au Japon (1889).
(29) Meteorological observations on the top of Fuji, 1890.
(30) Low pressure in Japan.

LIST OF PERSONS, PLACES, &c., TO WHICH THE PUBLICATIONS ARE SUPPLIED FREE OF COST.

Japan.
Bureau of the Archives.
Bureau of the " *Official Gazette.*"
Statistical Bureau.
Board of Imperial Chamberlain.
Board of the services to H. I. H. the Prince Imperial.
Bureau of the Imperial Estates.
Imperial Library.
Imperial Museum.
Noble's School for Boys.
Department for Foreign Affairs.
Japanese Legations in Foreign Countries.
Japanese Consulates in Foreign Countries.
Department for Home Affairs.
Civil Engineering Bureau.
Bureau for the Miscellaneous Affairs.
Central Sanitary Institute.
Special Capital's Section in the Department for Finance.
Medical Bureau in the Department for Wars.
Military College.
Land Survey Bureau.
College for Training Military Officers.
Naval College.
College for Training Naval Officers.
Central Sanitary Congress.
Gunpowder Manufactory in the Department for Wars.
Hydrographic Bureau.
Medical School for Navy.

Admiralty Stations.
College of Engineering in the Imperial University.
College of Science in the Imperial University.
College of Agriculture in the Imperial University.
Higher Normal School.
First Higher Middle School.
Second Higher Middle School.
Third Higher Middle School.
Fourth Higher Middle School.
Fifth Higher Middle School.
Yamaguchi Higher Middle School.
Kagoshima Higher Middle School (*Zoshikan*).
Higher Commercial School.
Tokio Industrial School.
Tokio Library.
Agricultural Bureau.
Forestry Bureau.
Geological Institute.
Commercial and Industrial Bureau.
Telegraph Bureau.
Postal Bureau.
Superintendence Office of Nautical Signals.
Mercantile Marine Office.
Tokio Navigation School.
Hakodate Navigation School.
Storm Signal Stations.
Yokohama West Wharf.
Shimbashi Railway Station.
Ueno Railway Station.
Yokohama Railway Station.
Takasaki Railway Station.
Provincial Meteorological Stations.
Erimo Lighthouse.
Mikomoto Lighthouse.
Fusan Telegraph and Post Office.
Prefectural Government Offices.
District Offices.
Japan Mail Steam Ship Company (*Usen Kaisha*).
Great Northern Telegraph Company, Nagasaki.
"*Japan Mail*" Office, Yokohama.
Minister President of State.
Minister of State for Home Affairs.

Minister of State for Finance.
Minister of State for Agriculture and Commerce.
Minister of State for Education.
Vice-Minister for Home Affairs.
Vice-Minister for Finance.
Vice-Minister for Agriculture and Commerce.
Vice-Minister for Education.

North America.
Harvard College, Cambridge, Massachusetts.
Hydrographic Office, U. S. A.
Mr. Lieutenant Finly, Chief of the Pacific Coast Weather Service, San Francisco.
Library, U. S. Weather Bureau, Department of Agriculture, Washington.
Director of the Blue Hill Meteorological Observatory.
Director of the Ohio Meteorological Bureau.
The State Board of Health, Michigan.
Meteorological Observatory, Central Park, New York.
Agricultural Experiment Stations, Lincoln, Nebraska.
Washburn Observatory, University of Wisconsin, U. S. A.
Iowa Weather Service, U. S. A.
Colegio Pio de Villa, Colon Panama, U. S. of Columbia.

South America.
Observatoire Impérial Astronomique et Météorologique, Rio Janeiro, Brésil.
Officina Central Meteorologica de Chili.
Observatorio Meteorologico del Colegio Pio de Villa Colon, Montevidio, Uruguay.
Academia Nacional de Ciencias, Cordoba, República Argentia.

Mexico.
Observatorio Meteorológico-Magnético Central.
Observatorio Meteorologico del Colegio del Estado de Puebla.
Sociedad Cientifica "Antonio Aizate"

Germany.
K. Preussisches Meteorologisches Institut, Berlin.
Dr. Carl Lang, Director der K. B. Meteorologische Centralstation, München.
Deutsche Seewarte, Hamburg.
Hydrographisches Amt des Reichs Marineamts, Berlin.
K. Sachsisches Meteorologisches Institut, Chemnitz.
Wetterwarte, Magdeburg.
K. Württenbergisches Meteorologisches Centralstation, Stuttgart.
Herr. E. Knipping, Hamburg.

England.

Meteorological Office, London.

Rousdon Observatory.

Editor of " *Meteorological Magazine*," London.

Royal Meteorological Society, London.

Kew Observatory, Richmond Surrey.

Radcliffe Observatory, Oxford.

Secretary of the Meteorological Council.

France.

Bureau Central Météorologique de France.

Société de Géographie, Paris.

M. le Directeur de l'Observatoire Météorologique de Pyrénées Orientales.

Directeur de l'Observatoire Météorologique de Lyon.

Observatoire Météorologique de Montsouris, Paris.

Observatoire de la Tour St.-Jacques, Paris.

Italy.

Ufficio Centrale di Meteorologia e Geodinamica, Roma.

R. Observatorio Astronomico di Brera, Milano.

Pontifica Universita Gregoriana, Roma.

Austria.

Centralanstalt für Meteorologie, Wien.

Centralanstalt für Meteorologie und Erdmagnetismus, Budapest, Ungarn.

Seewarte des Hydrographischen Amtes der K. K. Kriegsmarine zu Pola.

K. Oberrealschule zu Agram.

Observatorio Marittimo di Trieste.

Spain.

Observatorio de Madrid.

Sweden.

Observatoire Météorologique de l'Université, Upsala.

Norway.

Norwegischen Meteorologischen Institut, Christiania.

Denmark.

Dansk. Meteorologiska Institut, Copenhagen.

Roumania.

Institut Météorologique de Roumanie.

Switzerland.

Observatoire de Genève.

Schweizerische Meteorologischen Central-Anstalt, Zürich.

Belgium.

Observatoire royal de Uccle.

Holland.

K. N. Meteorologische Institut, Utrecht.

Africa.

Service Météorologique du Government Général de l'Algérie.

Royal Observatory, Cape of Good Hope.

Institut Egyptien, Le Caire.

India.

Meteorological Reporter of the Government of India, Calcutta.

Bombay Observatory.

Madras Observatory.

Russia.

Physikalisches Central-Observatorium, St. Petersburg.

Imperial Russian Observatory, Vladivostok.

Tifliser Physikalisches Observatorium, Tiflis.

Dr. H. Fritsche, Wassilji Ostrow, St. Petersburg.

Corea.

Corean Maritime Custom's Meteorological Service, Jenchuan.

Corean Maritime Custom's Meteorological Service, Fusan.

Corean Maritime Custom's Meteorological Service, Yuensan.

Philippine Islands.

Meteorological Observatory, Manila.

China.

Observatoire Météorologique, Zikawei.

Great Northern Telegraph Company, Shanghai.

Mr. W. Doberck, Government Astronomer, Hongkong Observatory.

Australia.

Melbourne Observatory, Victoria.

Sydney Observatory, New South Wales.

Adelaide Observatory.

Perth Observatory, West Australia.

Telegraph Department, Chief Weather Bureau, Brisbane, Queensland.

Java.

Observatorium te Batavia, Nederlandsch-Indië.

Canada.

Meteorological Office, Toronto.

ADMINISTRATION AND INSPECTION.

The management of business, so far as it is not of a technical nature, is entrusted to subordinate officials under the superintendence of the Director who generally undertakes the correspondence with any of the Government Offices and with foreign countries.

Once a week the Director assembles all the meteorologists and other officials belonging to each Service in order to discuss questions specially relating to the business.

The Library of the Observatory contains 1,739 volumes of which

557	are	Japanese,
842		English,
282		French,
58		German,

and a great number of miscellaneous scientific periodicals.

The expenditure of the Central Meteorological Observatory is at present as follows :—

Salaries	*yen* 9,280.00
Miscellaneous expenses	7,080.75
Total........... .. ;......................	16,360.75

The expenditure for the Provincial Stations:—

1st order {
Salaries	*yen* 581	in average
Miscellaneous expenses	752	
Total	1,333	

2nd order {
Salaries	*yen* 326	in average
Miscellaneous expenses	760	
Total	1,086	

Inspections of the Provincial Stations are made by Titular Meteorologists of the Central Meteorological Observatory.

There is no special inspector, but the sectional-chief are sent to inspect the Stations; generally each Station is visited by a Titular Meteorologist or Assistant Meteorologist at regular intervals.

The inspector compares the instruments used there with his standard, and the arrangements, books, etc., are subjected to a thorough revision.

HISTORICAL SUMMARY OF THE SYSTEM.

1872, July:	Hakoda temeteorological station was founded and opened tri-daily observations.
1875, 1st June:	A meteorological service was established in Land Survey Section of Geographical Bureau in Home Department, and opened tri-daily observations. The Chief Commissioner of the Bureau was Sugiura Yuzuru and all meteorological matters have since been entrusted to H. B. Joyner, an English employé.
August:	Geographical Bureau collected meteorological records from all lighthouses in Japan.
November:	Report of meteorological observations have been first printed in the "*Japan Daily Mail*" (an English newspaper published at Yokohama).
December:	In the "*Tokio Nichi Nichi Shimbun*" (Tokio daily press).
1876, 1st January:	Meteorological observations were commenced in Kobe.
March:	Every 5 day, monthly and half-yearly reports were published in Geographical Bureau.
May:	H. B. Joyner presented to the Government his view as to the Stormsignals.
1st Sept.:	Sapporo station was opened.
1877, April:	Meteorological reports have been exchanged with Chief Signal Office, Washington, U. S. A.
30th June:	H. B. Joyner was released on expiration of the term of engagement.
24th Sept.:	T. Sakurai was appointed the Chief Commissioner of Geographical Bureau.
1878, January:	Meteorological reports have been exchanged with Mexico and other countries.
February:	The reports of international simultaneous observations have been forwarded to Chief Signal Office, Washington, U. S. A.
1st June:	Observations of earth thermometer have been commenced.
1st July:	Nagasaki station was erected and tri-daily observations have been commenced since.
17th July:	Evaporometer was supplied to the Observatory.
1st Sept.:	Rumoye station was erected.
1879, January:	Photographic hygrograph was used in the Observatory and Hiroshima station opened.
March:	Palmieri's Seismograph was supplied to the Observatory.
2nd June:	Shinagawa Yajiro was appointed the post of the Chief of Geogra-

phical Bureau.

1st July :	Wakayama and Nemuro stations were established.
July :	Meteorological observations have since been exchanged with Berlin, Peking and Manila observatories.
1880, 5th March :	T. Sakurai resumed the post of the Chief Commissioner of Geographical Bureau.
15th Oct. :	Kioto station was opened.
1881, 1st April :	Rumoye station was closed and Mashike was opened.
1st July :	Nobiru and Niigata stations were established.
October :	Meteorological observations have since been exchanged with Belgium and 8 other observatories.
1882, 1st January :	Aomori and Kanazawa stations were opened.
January :	The service of weather forecast and storm warning was proposed by and entrusted to a German employé, E. Knipping.
1st March :	Kochi station was opened.
1st July :	The Tokio Meteorological Observatory in Akasaka was removed to Hommaru within the precinct of the Imperial Castle Tokio, Ozaka station opened, and metric system has been adopted since.
1st August :	Simultaneous observations at 6^h am, 2^h pm, and 10^h pm (Kioto time) daily were commenced in the Tokio Observatory and Provincial Stations, in addition to the regular ones.
September :	"Instructions to observers" were printed and distributed.
1st October :	Akita station was opened.
November :	An application was made to the Government for permission to collect once a day meteorological telegrams from all the Provincial Stations, free of charge.
1st Dec. :	Numazu and Hamamatsu stations were opened.
1883, 1st January :	Sakai, Shimonoseki, Kagoshima, Miyazaki stations were opened ; the Government granted the permission to the Observatory to collect meteorological telegrams from the Stations free of charge, the special arrangement of free postage was made.
1st Feb. :	Gifu station was opened.
16th Feb. :	A system of weather telegraphy was inaugurated and weathermaps first printed.
1st March :	Miyako station was founded.
26th May :	Storm warnings were first issued.
19th July :	The instructions regarding the storm warnings and stormsignals were published in the "Official Gazette".
August :	Milne-Gray's Seismograph was supplied to the Observatory.
1st Nov. :	A signal station was erected at Shinagawa.
1st Dec. :	An application was made to the Government for permission to collect meteorological telegrams thrice a day, free of charge.

1884, 1st April:	Saga and Suttsu stations were erected.
10th May:	The permission above mentioned was obtained.
16th June:	Meteorological observations have since been made at the Imperial Japanese Post Office, Fusan.
June:	Weather forecast has been posted at every police station in Tokio, and weathermaps published three times a day.
27th Sept.:	Seismological service was organized.
November:	Meteorological observations have been commenced in Bonin Island.
December:	Saga station was closed.
1885, 1st Feb.:	Fushiki station was founded by Fushiki residents.
25th June:	Survey Section was closed and a Fourth Section established in Geographical Bureau, in which the matters concerning climatology, weather telegraphy and seismology were dealt with.
1st Nov.:	Mashike station was closed and Soya station opened.
1886, 1st January:	Hourly observations have been commenced in Tokio.
19th Jan.:	The Fourth Section having been closed, Observation Section was established, in which the business concerning climatology, weather telegraphy, magnetism and seismology have been transacted.
20th Jan.:	K. Kobayashi was appointed the sectional-chief.
1st Sept.:	Choshi station was opened by the people of that place, on the same date Erimo station.
15th Sept.:	Izugahara station was opened.
1887, January:	Tokio Meteorological Observatory has been termed "Central Meteorological Observatory"; Oita station was opened in this month.
3rd August:	The Imperial Ordinance relating to the Central Meteorological Observatory and Meteorological Station was promulgated.
1st Sept.:	Nobiru station was closed and Ishinomaki opened.
1st October:	The positions of 51 Provincial Stations were notified to the public.
1888, 1st January:	Japan time has since been used throughout the Empire.
10th March:	Weather forecasts for 24 hours in advance have been printed in the "*Official Gazette*" and many Tokio newspapers.
1st April:	Weathermaps have been printed once a day in one sheet.
1st Nov.:	The first meteorological meeting was held in the Central Meteorological Observatory, at which all the chiefs or observers in charge of Provincial Stations attended.
1889, 1st May:	Fukushima station was opened.
1st July:	Yamagata and Tsu stations were established.
7th August:	I. Arai was appointed the Director of the Central Meteorological Observatory.
1st Dec.:	Kushiro station was founded.
1890, 31st March:	I. Arai, the Director, was put on the Retired List and K. Koba-

yashi was made acting director; E. Knipping returned to Germany on expiration of his term of engagement.

1st July: Nagoya and Naha (Liukiu) stations were founded.

1st August: Saga station re-opened.

2nd August: The staff organization of the Observatory promulgated.

15th August: Utsunomiya station opened.

15th Nov.: Okayama station was established.

1891, 1st January: A system of coast warning has been opened and temperature indication issued on the events of remarkable changes.

1st April: Tokushima station was opened.

1st June: The time of issuing indications was changed from 9^h pm to 4^h pm, weathermaps hitherto printed at about 8^h am have been issued since the above date at about 4^h pm.

10th June: Instructions were given from the Home Department to 8 Prefectures, viz. Kanagawa, Hiogo, Saitama, Gumma, Ibaraki, Yamanashi, Shiga and Shimane, that they should establish, if possible, all the Provincial Stations before the end of the 26th fiscal year (1893).

20th June: The chiefs or observers in charge of all Provincial Stations were assembled to the Central Meteorological Observatory where the second meteorological meeting was held for 10 days.

1st July: New Japanese ideographs corresponding to the French terms of the metric system have since been used in the Observatory and Stations.

24th July: Amendment in the Organization of the Observatory was promulgated.

16th August: K. Kobayashi was appointed the Director of the Central Meteorological Observatory.

1892, 1st January: The time of despatching regular meteorological telegrams have been changed from 9^h pm to 10^h pm; Tokachi (in Hokkaido) and Tonno (Instructor of the Yamaguchi Higher Middle School) private Yamaguchi stations were opened.

5th May: The Rules for enforcement of the Regulations relating to the Meteorological Observatory and Provincial Stations were revised and published.

15th June: The revised Rules regarding Meteorological Telegrams have been put in force, and the Rules relating to Meteorological Signals, Instructions regarding the Provincial Indications and Terms to be used in indications were approved; Tokio weather forecast was first issued.